FIRE SAFETY

Printed in the United States of America.

Library of Congress Cataloging-in-Publication Data
Loewen, Nancy, 1964-
.Fire Safety/Nancy Loewen
p.   cm.
Includes index.
Summary: Shows what to do and what not to do to practice fire safety sense.
ISBN 1-56766-258-7   (Lib. bdg.)
1. Fire prevention--Juvenile literature.   [1. Fire prevention.   2. Safety.]
I. Title.
TH9148.L63   1996
628.9'2--dc20   95-25896
CIP
AC

FIRE SAFETY

By Nancy Loewen Illustrated by Penny Dann

THE CHILD'S WORLD

Lots of people think fire is **fascinating**. It gives us heat and light and it can be used to roast marshmallows, too! But fire can also be dangerous, even deadly. That's why it's important to learn how to use fire the right way, and to prevent accidental fires. Pickles and Roy will show you what to do—and what not to do—to practice fire safety sense!

Don't play with matches, lighters, candles, or anything else that could start a fire.

Keep books, paper, and clothing at least three feet away from space heaters, irons, toasters and other hot **appliances**.

Never use the stove or oven unless an adult is with you. Pay close attention to what you are doing.

Always keep a fire extinguisher in the kitchen. That way, if a small fire does start, it can be put out quickly. If it can't be put out easily, though, don't waste any time. Get out and call for help!

If you have a fireplace or wood-burning stove, you need to learn some special rules.

Never build a fire by yourself—an adult should do that task.

Use a sturdy metal screen around the fireplace to
keep sparks from flying out.

Remind your parents to have
the chimney cleaned at least
once a year.

Sometimes fires can start no matter how careful you are. But there are many things you can do to reduce your risk of injury.

Smoke detectors are a very important safety tool. They will warn you when there is a fire. Test the smoke detectors at least once a month, to make sure they are working.

Develop a fire escape plan with
your family. Learn two safe
ways to exit out of every room
such as through a window,
door or fire escape.

If you live in an apartment, know the stairway and fire exit locations.

Pick a meeting spot in front of your house or apartment building. That way it will be easy to account for everyone.

Choose a family member to call 911 from a neighbor's home.

Practice fire drills twice a year, to make sure everyone in your household knows what to do if there's a fire.

Sleep with the bedroom doors closed to keep out smoke and flames. If you awaken to the smell of smoke or the smoke detector goes off, shout to **alert** your family. The safest air is close to the ground, so crawl to your door. Feel the door. If it is hot, don't open it! If the bedroom door is cool, open it slowly. Crawl on your hands and knees to the nearest exit.

If you can't exit through the door, try to escape through the window. If you can't escape out the window, stuff the space under the door with clothes or towels. Then stay by the window to **signal** to people outside. Don't break the window or jump down unless you have no other choice.

Don't waste time by trying to find your pets or toys.
It's more important that you get out safely.

Once you're out, stay out! Never go back into a burning building, even if a family member is still inside. Leave the rescuing to the fire fighters, who are trained and have the right equipment.

If a fire alarm sounds when you're in a public building, look for the nearest exit sign and go there. Walk, don't run. If you fall and hurt yourself, you'll have a harder time getting out.

Never use the elevator during a fire. It could break and get stuck between floors, or go straight to the fire! Use the stairs instead—no matter how many flights you have to walk.

If you ever catch on fire yourself, don't run! That will only make the fire worse. Drop immediately to the ground and start rolling around to put out the flames. Cover your face with your hands. Remember: stop, drop, and roll!

Learning about fire safety might seem like a big responsibility. And it is! But knowing what to do will make it less likely that you'll ever be hurt in a fire. Ask your parents, teachers, and friends to help you learn the fire safety rules. Keep fire in its place!

23

Glossary

alert (A-lert)
to make others aware of danger. If you smell smoke, shout to alet your family.

appliances (a-PLY-an-says)
a household or office device. Keep clothing away from hot appliances.

fascinating (FAS-si-nate-ing)
very interesting or charming. Lots of people think fire is fascinating.

signal (CIG-null)
to communicate through an action or gesture. If you can't escape a burning building, stay by the window and signal the people outside.